ISBN 978-1-334-30741-6
PIBN 10657757

This book is a reproduction of an important historical work. Forgotten Books uses state-of-the-art technology to digitally reconstruct the work, preserving the original format whilst repairing imperfections present in the aged copy. In rare cases, an imperfection in the original, such as a blemish or missing page, may be replicated in our edition. We do, however, repair the vast majority of imperfections successfully; any imperfections that remain are intentionally left to preserve the state of such historical works.

1 MONTH OF
FREE
READING

at
www.ForgottenBooks.com

By purchasing this book you are
eligible for one month membership to
ForgottenBooks.com, giving you
unlimited access to our entire
collection of over 700,000 titles via
our web site and mobile apps.

To claim your free month visit:
www.forgottenbooks.com/free657757

English
Français
Deutsche
Italiano
Español
Português

www.forgottenbooks.com

Mythology Photography **Fiction**
Fishing Christianity **Art** Cooking
Essays Buddhism Freemasonry
Medicine **Biology** Music **Ancient
Egypt** Evolution Carpentry Physics
Dance Geology **Mathematics** Fitness
Shakespeare **Folklore** Yoga Marketing
Confidence Immortality Biographies
Poetry **Psychology** Witchcraft
Electronics Chemistry History **Law**
Accounting **Philosophy** Anthropology
Alchemy Drama Quantum Mechanics
Atheism Sexual Health **Ancient History**
Entrepreneurship Languages Sport
Paleontology Needlework Islam
Metaphysics Investment Archaeology
Parenting Statistics Criminology
Motivational

THE

FORESTS OF CANADA.

BY

ROBERT BELL, B.A.Sc., M.D., LL.D.,

Assistant Director of the Geological Survey, Ottawa.

ıprinted from the RECORD OF SCIENCE, Vol. II., No. 2, 1886.)

MONTREAL :

GAZETTE PRINTING COMPANY.

1886.

THE

FORESTS OF CANADA.

BY

ROBERT BELL, B.A.Sc., M.D., LL.D.,

Assistant Director of the Geological Survey, Ottawa.

(Reprinted from the RECORD OF SCIENCE, Vol. II., No. 2, 1886.)

MONTREAL :

GAZETTE PRINTING COMPANY.

1886.

BY

ROBERT BELL, B.A.Sc., M.D., LL.D.,

Assistant Director of the Geological Survey, Ottawa

—————— ♦ ——————

Printed from the RECORD OF SCIENCE, Vol. II, No. 2, 1880.

—————— ♦ ——————

MONTREAL.

GAZETTE PRINTING COMPANY.

1880.

erty of their soil, will never be cultivated to any great
extent. This great coniferous belt has a crescentic form,
curving southward from Labrador to the far Northwest,
keeping Hudson Bay on its northern side. The distribution
of our forests appears to be governed almost entirely by
existing climatic conditions, although it may be modified to
some extent by the geological character of different districts;
and some of the peculiarities of their present distribution
may be due to former conditions affecting their dispersion.

Beyond the northern limit of the forests on the mainland
of the continent there is a large triangular area to the northwest and another to the north-east of Hudson Bay, called
the Barren Grounds, which are destitute of trees solely on
acçount of the severity of the climate, as the other conditions do not appear to differ from those of the adjacent
wooded regions to the south. The treeless region of the
Western States and the south-western part of the Northwest Territories of Canada are called plains as distinguished
from the prairies, which often are partially wooded. The
latter occupy an immense space between the plains and the
forest regions to the east and north-east. The plain and
prairie conditions are also due to climate, and not, as some
have supposed, to fires having swept away formerly existing
forests. This is shown by the contours of the lines marking
the western limits of the various kinds of trees which pre-
vail in the east, as well as from the absence of water-courses,
which would exist if sufficient rain had fallen in comparatively recent times to have maintained forest growth.

Although the Dominion embraces about half of the continent, only some ninety out of the 340 species of the forest
trees of North America are found within her borders,
including the Pacific slope. Yet the area under timber in
Canada is perhaps as great as that in the United States.
It is, therefore, evident that the forests are less diversified
in the north than in the south. This is in accordance with
the general law of the greater richness of the flora of warm
countries; but it may be due also in part to the fact that in
the north we have greater uniformity of physical and
climatic conditions over wide areas than in the south. For

example, we have a similarity in these conditions from Newfoundland to Alaska, and hence throughout the great distance of 4,000 miles we find the same group of trees. Again in the great triangular area of the Northwest, between the United States boundary, the Rocky Mountains, and the Laurentian region, embracing over 600,000 square miles, very little difference could be observed in the climate, the soil, or the general level of the country, and hence the same group of trees—only about half a dozen in all—is found throughout this immense tract. In striking contrast with this is the fact that on the same farm lot in the south-western part of the Province of Ontario one may often count as many as fifty different kinds of trees. The richness in variety of the native trees of Ontario and the adjacent States is owing to the fertile soil and the favourable conditions as to summer temperature, constant moisture, and the absence of intense cold in the winter.

The writer exhibited a map showing the northern and western limits of the principal forest trees of the Dominion east of the Rocky Mountains. From this it appears that the range of species is not according to the mean annual temperature or precipitation, but rather to the absence of extremes of heat and cold, and of great dryness. For these reasons a number of the trees of the Province of Quebec and northern Ontario do not range west into Manitoba, although the annual means of temperature and precipitation are nearly the same in both. This map also shows in a striking manner that the northern limits of our various forest trees are by no means parallel to one another, although locally some groups may be nearly so for a certain distance. Some of them pursue extraordinary or eccentric courses, which are difficult to account for. The most remarkable of these is the white cedar, which in the central part of its trend reaches James Bay, but drops suddenly to the south at the Gulf of St. Lawrence in the east, and on reaching the longitude of the head of Lake Superior in the west. Yet the climate and other conditions appear to be the same for some distance both east and west of these lateral boundaries. An outlying colony of the white cedar is found at Cedar Lake near the north-western part of Lake Winnipeg. Colonies or outlying patches of other trees have been noted in different localities, such as of the basswood and sugar-maple at Lake St. John, north of Quebec, of the grey elm on Missinaibi River, near James Bay, and of the hemlock spruce at Thompson, near the west end of Lake Superior.

Rivers and lakes, by supplying heat and moisture and warding off summer frosts, often promote the growth of trees on their immediate banks which are not found elsewhere in the surrounding country. Instances of this may be seen along the North Saskatchewan, where the negundo, green ash, grey elm, white birch, alder, etc., thrive only on the river banks. In the cold regions, the white spruce grows to a much larger size on the shores and islands of rivers flowing north than elsewhere. It has been found that exotic fruit trees and other introduced plants can be successfully cultivated around the shores of the larger lakes, especially on their southern sides, which will not grow at a short distance inland. On the other hand, the immediate proximity of the sea, with a lower summer temperature than the land, is unfavourable to the growth of timber in the north. The habits of some trees are much modified in different latitudes. Species which grow in warm dry soil in the north may be found in cold, heavy, or wet land in the south. The larch, balsam, white cedar, white pine, white birch, etc., are examples of this tendency. Some species extend far to the south of their general home along mountain ridges, while others seem to refuse to follow such lines. The existence of extensive swamps, the shelter of hills, or the elevations which they afford, are therefore to be regarded as among the minor conditions governing the distribution of trees.

The peculiarities in the outline of the northward limit of the white cedar and other species of trees, may throw some light on questions as to the direction from which they have migrated or been dispersed. In some cases which the author has studied, the trees appear to have reached the most northern limit possible. For example, in its most northern range,

the first tender leaves and shoots of the black ash are
blighted almost every year by the spring frosts; the trees
are of small size or stunted in height, and only occasionally
bear seed. Sir John Richardson mentions that, on the barren
grounds, outlying patches of dying spruces were sometimes
met with far out from the verge of the main forest, and that
he saw no evidence of young trees springing up beyond the
general line of trees; from which he infers that the latter
is retreating southward. A similar condition is said to
exist in Siberia.

In tracing the northern limits of several of the trees as
laid down on the author's map, it would be observed that
the northward variations from the general direction usually
corresponded with depressions in the country, while the
southward curves occurred where the elevations were
greatest. The height-of-land dividing the waters of the St.
Lawrence from those of Hudson Bay has a general paral-
lelism with the northern limits of many of the species; but
as the watershed is not marked by any great elevation or
by a ridge, the circumstance referred to may be owing
simply to the accident of its trend coinciding with the aver-
age course of the isothermal lines.

The author divides the trees of the Dominion east of the
Rocky Mountains into four groups in regard to geographi-
cal distribution, namely: (1) A northern group, including
the white and black spruces, larch, Banksian pine, balsam-
fir, aspen, balsam-poplar, canoe birch, willows and alder,—
these cover the vast territory from the northern edge of the
forests down to about the line at which the white pine
begins; (2) a central group of about forty species, occupy-
ing the belt of country from the white-pine line to that of
the button-wood; (3) a southern group, embracing the
button-wood, black walnut, the hickories, chestnut, tulip
tree, prickly ash, sassafras, and flowering dog-wood, which
are found only in a small area in the southern part of
Ontario; (4) a western group, consisting of the ash-leaved
maple, bur-oak, cotton-wood, and green ash, which are
scattered sparingly over the prairie and partially-wooded
regions west of Red River and Lake Winnipeg.

The finest timber of the second group within the limits of Canada is to be met with along the east side of Lake Huron in the counties of Lambton, Huron and Bruce, where the button-wood, elm, maple, yellow birch, cherry, bass-wood and hemlock attain a height of one hundred feet and upwards. Although the Ottawa valley has produced more white pine timber than any other region in the Dominion, the largest and finest trees grow on the sandy soils of the counties bordering the northern sides of Lake Erie and of the western part of Lake Ontario, where extensive and splendid pineries stood when these regions were first invaded by the white man. In the Northwest Territories, the largest trees are the elms along the rivers (which, however, do not extend far north) and the rough-barked poplars, which, even as far north as the Laird and the lower Mackenzie, have trunks five feet in diameter. Along Athabasca River the author had seen spruces which measured ten and twelve feet in girth.

The distribution of our forest trees affords us one of the most obvious tests of climate, and although it may not be more reliable than that of the smaller plants, it is more noticeable by the common observer. In the older provinces of Canada the settlers are often guided to a great extent in their selection of land by the kinds of trees it supports, a thrifty growth of beech and sugar-maple, for instance, being generally considered a good sign ; but such tests must necessarily be only of local application. In the prairie region, timber may be entirely absent from the finest soil, while the least hardy trees of the west flourish in the stiff clay-banks or among the stones along the rivers on account of the moisture and heat derived from the water.

The map which has been referred to is useful in defining the extent of country over which each kind of timber was to be found. But in estimating the quantities which may be yet available for commercial purposes in the regions still untouched by man, various circumstances require to be considered, such as the favourable or unfavourable conditions of soil, etc., as well as the proportion which has been destroyed by fire, and other causes. The amount of timber which has been lost through forest fires in Canada is almost incredible, and can only be appreciated by those who have travelled much in our northern districts. The proportion of white and red pine which has been thus swept away in the Ottawa Valley and in the St. Maurice and Georgian Bay regions, is estimated by the lumbermen as many times greater than all that has been cut by the axe. Yet all this is insignificant in quantity compared with the pine, spruce, cedar, larch, balsam, etc., which has been destroyed by this means in the more northern latitudes all the way from the Gulf of St. Lawrence to Nelson River, and thence north-westward. It is true that the commercial value of this timber was not so great as that of the more southern pine regions which have also been partially ruined. The total quantities which have disappeared are almost incalculable, but even a rough estimate of the amount for each hundred or thousand square miles shows it to have been enormous, and of serious national consequence. The writer had traversed these great regions in many directions, and could testify to the widespread devastation which had taken place. Nearly every district was more or less burnt, the portions which had been overrun by fire usually exceeding those which remained green. These northern coniferous forests were more liable than others to be thus destroyed. In the summer weather, when their gummy tops and the mossy ground are alike dry, they burn with almost explosive rapidity. Small trees are thickly mingled with the larger ones, and they all stand so closely together that their compact branches touch each other, thus forming a sufficient-ly dense fuel to support a continuous sheet of flame on a grand scale. Before a high wind the fire sweeps on with a roaring noise, and at a rate which prevents the birds and beasts from escaping. Thus, in one day, the appearance which a large tract of country is to wear for a hundred years may be completely altered. After a time the burnt district becomes overgrown, first with shrubs and bushes, then with aspens and white birches, among which coniferous trees by-and-by appear; but finally at the end of a hundred and fifty years or more they regain possession of the burnt tract. This process of alternation of crops of timber appears to have been going on for centuries, but in modern times the fires must have been more numerous and frequent than formerly.

Along Moose River and the lower part of the Missinaibi, the original dark coniferous forest of these latitudes is replaced by the light green poplars and white birches, for more than a hundred miles, and this condition has existed since the memory of the oldest Indian of the district. Here and there may be seen a patch of large spruce—remnants of the original forest—and everywhere under the deciduous growth, the charred stumps of the old conifers may be found. On the east side of the southern part of Lake Winnipeg, and nearly all along Winnipeg River, the principal forests have been destroyed by fire, and replaced by aspen and white birch.

Forest fires are undoubtedly due occasionally to lightning, the author having once actually witnessed the origin of a fire in this way, and he had often been informed by the Indians that they had seen similar cases. But most of them are traceable to the carelessness of white men and demoralized Indians. In the partially inhabited regions, most of the forest fires originate by the settlers burning brush and log-heaps in clearing the land. It may be asked if we have no means of stopping this fearful destruction of the timber of the country. Laws on the subject do exist, but no adequate means appear to be provided for enforcing them. The author recommended a reform in this respect, before it be too late. Crown lands of real value for agriculture should be separated for the purpose of administration from those which are acknowledged to be useful only or principally for their timber, and settlement should be prohibited within the latter. Heretofore, the great consideration of Government was the peopling of the country, the timber being looked upon as of secondary importance, and it was willingly sacrificed in the interests of the settler, who came to regard it as his natural enemy. The time has come when we must change all this. In the absence of forest guardians and proper regulations, lumbermen have often to submit to a species of blackmail from discharged employees and pretending settlers in order to keep them off their limits. Indians sometimes burn the forests off each other's hunting-grounds from motives of revenge, but as a rule the fires which they start are from carelessness or indifference. When cautioned in a friendly way, they are willing to exercise greater care, and the beneficial effects of this course are already manifest in the region between Lake Winnipeg and Hudson Bay, where the author had remonstrated with them on the subject. He suggests that the annuities which they receive from Government be withheld as a punishment for burning the woods, or that a bounty be paid each year that no fires occur. In this way the Indian chiefs and headmen may be made the most efficient and earnest forest guardians we could possibly have.

Fires are not so liable to run in forests of full-grown white and red pines, such as those of southern Ontario, which have suffered comparatively little from this cause, but have now been mostly cut down and utilized by the lumbermen. Hardwood forests are seldom burnt to any great extent, except where the soil is shallow and becomes parched in summer, as, for instance, on the flat limestone rocks of Grand Manitoulin Island and the Indian peninsula of Lake Huron, through much of which fires have run, burning the vegetable mould and killing the roots, thus causing the trees to fall over even before they have decayed. Hence the term "fire-falls" applied in such cases.

If we had educated and intelligent conservators of forests in Canada, appointed by the Government, their duties, in addition to preventing the destruction of the timber by fire and otherwise, might be directed to promoting the growth of existing timber, encouraging transplanting, the introduction of foreign trees which might grow in this country, the dissemination of information on practical forestry, etc., investigating the causes of diseases among trees, directing the attention of foreign purchasers to our woods and pointing out to our lumbermen possible new markets for timber products and for varieties of woods not now utilized. That disease does sometimes cause great havoc among our forests is illustrated by the recent fact that the spruces in New Brunswick, the principal timber tree of that province, died over extensive areas, a few years ago, and the disease has now spread into the Gaspé peninsula. It is supposed to be due to a fungus which attacks the roots, but it is not certain that the fungus itself may not be induced by the pre-existence of some other disease. In the Province of Quebec the larches or tamaracs, have sometimes died from unexplained causes in extensive tracts. As soon as coniferous trees have become scorched by fire or show signs of failing vitality, their trunks are attacked by boring beetles, and they must be immediately cut down and immersed in water if the timber is to be saved.

In regard to the future supplies of timber which may be available in Canada, the greater part of the white oak and rock elm has been already exported. The cherry, black walnut, red cedar, and hickory have likewise been practically exhausted. Red oak, basswood, white ash, white cedar, hemlock, butternut, hard maple, etc., as well as many inferior woods, are still to be found in sufficient quantity for home consumption. A considerable supply of yellow birch still exists, and in some regions it is yet almost untouched. Until recently there was an indistinct popular notion that the white pine, our great timber tree, extended thoughout a vast area in the northern parts of the Dominion, from which we might draw a supply for almost all time. The author's map showed, however, that its range was comparatively limited. The shaded portions of the accompanying little map will serve to give an idea of the extent of our pine lands, relatively to that of the whole Dominion. Even if we include the Douglas pine area of British Columbia, it will be seen to be small in comparison with the rest of Canada. And it must be observed that this shading represents the botanical and not the commercial distribution of the pine, and that the valuable timber has been already cut away or is very sparsely distributed through a large proportion of it. Although it was found over a very extensive district to the north-west of Lake Superior, it was very thinly scattered, of smaller size, and poorer quality than further south. Our principal reserves of white pine, as yet almost untouched, are to be found in the region around Lake Temiscaming, and thence westward to the eastern shore of Lake Superior, and in the

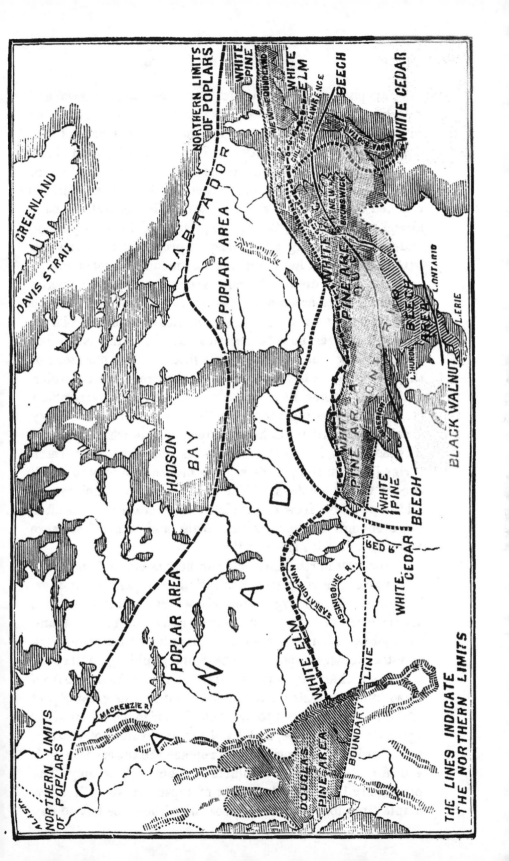

central parts of the district between the Ottawa and Georgian Bay. There is also more or less red pine in the district referred to. The newly constructed Canadian Pacific Railway between Lake Nipissing and Lake Superior has afforded a means of access to the centre of this great pine region, which could not so well be reached by any of the rivers. Lumbering operations have already begun near the railway west of Lake Nipissing, and unless the charges for transport prove too high, the probabilities are that hereafter a large amount of timber will be sent out of this district by rail. When the exportable white pine shall have become exhausted, as it must before many more years, we have still vast quantities of spruce and larch, which may even now be regarded as the principal timber available for this purpose in the future. But our stock of these woods is to be found mostly in the great country which drains into James Bay, whose numerous large rivers afford facilities for floating timber to the sea, and in the country thence westward to Lake Winnipeg. Fine white spruce is likwise found in some localities in the Northwest Territories between the prairie regions and the country of small timber to the north-east. The Banksian pine, which ranges all the way from New Brunswick to Mackenzie River, is often large enough for sawing into deals, and will afford large quantities of good railway ties.

If the vast northern forests can be preserved from fire in the future, our supply of small timber is practicably inexhaustible. When larger trees elsewhere shall have become scarce, much of it may some day be sawn into boards, scantling, joists, rafters, flooring, etc. Supplies of timber for railway-ties, telegraph-poles, mines, fencing, piling, small spars, cordwood, charcoal, paper-making, etc., may be drawn from these immense districts for all time, since the greater part of the regions referred to are not likely to be required for agricultural purposes, and by a proper system of cutting, a new growth will spring up to replace the timber removed, and in its turn become available to keep up the supply. The practically interminable extent of these forests will allow ample time for the smaller trees, which may be left on any ground cut over, to come to maturity before it is again called upon to furnish its quota. Some of the woods of the more southern districts of Canada, which have had little value hitherto, except for fuel, only require to be better known to be utilized for many purposes.

The people of Canada have heretofore been accustomed to such an abundance of wood, and to the idea that trees stood in the way of the progress of the country, that tree-planting has as yet made but little progress among us. A beginning has, however, been made in the last two years in the provinces of New Brunswick and Quebec, where "Arbor Days" have been proclaimed. In Ontario an Act was passed in 1883, and a fund set apart for the encouraging of tree-planting along highways. The time has arrived for more vigorous action by the general Government and the Local Legislatures looking to the improvement and preservation of the forests which still remain in Canada, and for the partial restoration of those which have been destroyed.

CPSIA information can be obtained
at www.ICGtesting.com
Printed in the USA
LVHW021322071118
596294LV00005B/856